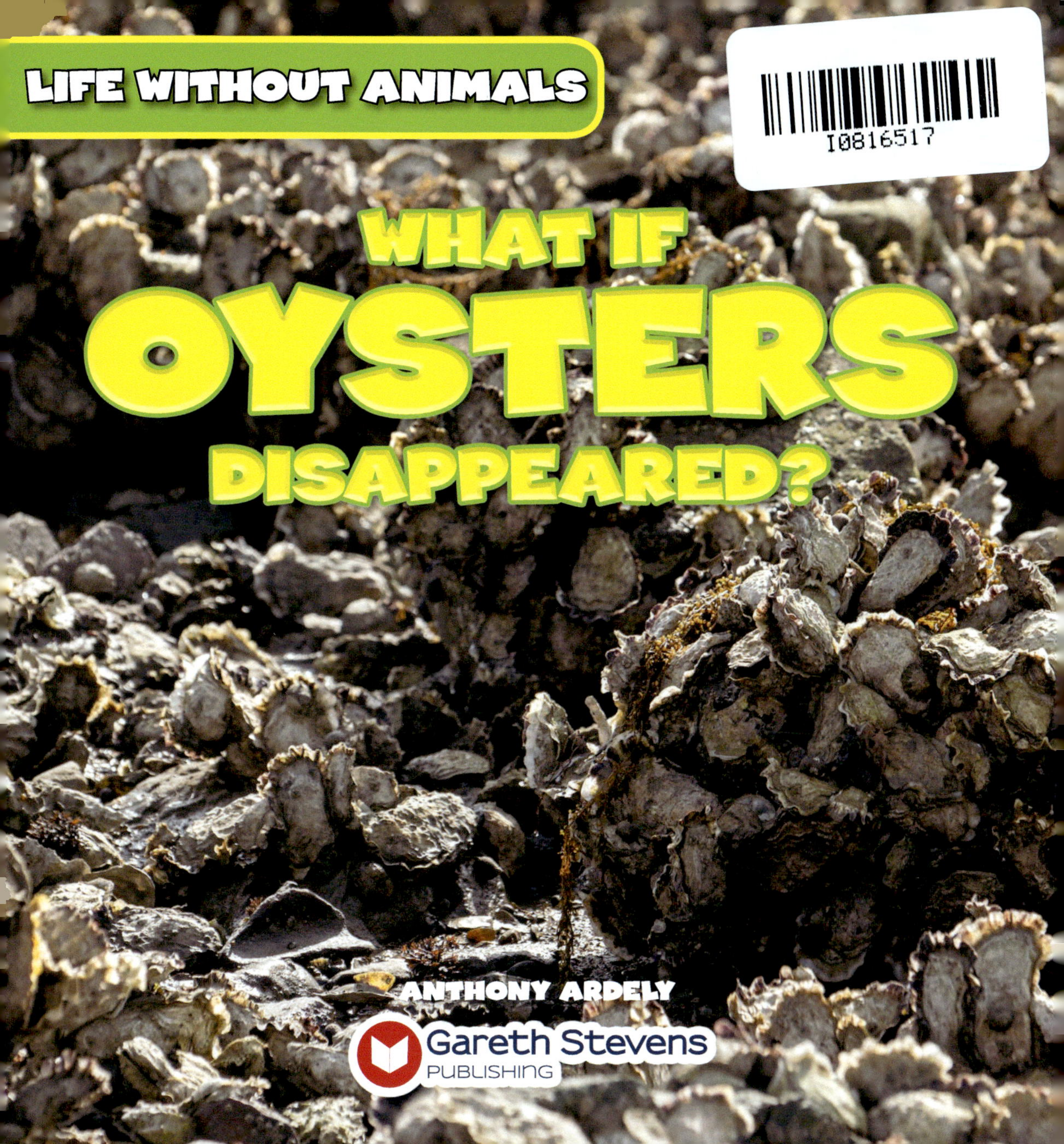
LIFE WITHOUT ANIMALS
I0816517
WHAT IF
OYSTERS
DISAPPEARED?
ANTHONY ARDELY
Gareth Stevens
PUBLISHING

Please visit our website, www.garethstevens.com. For a free color catalog of all our high-quality books, call toll free 1-800-542-2595 or fax 1-877-542-2596.

Library of Congress Cataloging-in-Publication Data

Names: Ardely, Anthony, author.
Title: What if oysters disappeared? / Anthony Ardely.
Description: New York : Gareth Stevens Publishing, [2023] | Series: Life without animals | Includes index.
Identifiers: LCCN 2021046498 | ISBN 9781538276365 (set) | ISBN 9781538276372 (library binding) | ISBN 9781538276358 (paperback) | ISBN 9781538276389 (ebook)
Subjects: LCSH: Oysters–Juvenile literature. | Oysters–Behavior–Juvenile literature. | Oysters–Conservation–Juvenile literature.
Classification: LCC QL430.7.O9 A73 2023 | DDC 594/.4–dc23
LC record available at https://lccn.loc.gov/2021046498

Published in 2023 by
Gareth Stevens Publishing
111 East 14th Street, Suite 349
New York, NY 10003

Designer: Rachel Rising
Editor: Kate Mikoley

Photo credits: Cover, pp. 1, 19 barmalini/Shutterstock.com; p. 5 Leonie Broekstra/Shutterstock.com; p. 7 golfza.357/Shutterstock.com; p. 9 foryouinf/Shutterstock.com; p. 11 papa studio/Shutterstock.com; p. 13 Danita Delimont/Shutterstock.com; p. 15 SJ Duran/Shutterstock.com; p. 17 ritthikorn50/Shutterstock.com; p. 21 flrwnx/Shutterstock.com.

Printed in the United States of America

CPSIA compliance information: Batch #CSGS23: For further information contact Gareth Stevens, New York, New York at 1-800-542-2595.

CONTENTS

All About Oysters 4
Rocks and Reefs 6
All Kinds of Oysters 8
People and Oysters 10
Home for Many 12
Keeping Things Clean 14
Stopping the Storm 16
We Need Oysters! 18
Helping Oysters 20
Glossary 22
For More Infomation 23
Index 24

Boldface words appear in the glossary.

All About Oysters

Oysters can be found in water along ocean shores. They are part of a group of animals called mollusks. Mollusks have soft bodies and no **backbones**. Like many other mollusks, oysters' bodies are covered by a hard shell.

Rocks and Reefs

Oysters commonly **attach** to rocks, shells, or other hard places below the water. They may even attach to other oysters' shells, forming **colonies**. When many oysters attach together, they form a chain of shells known as an oyster reef.

All Kinds of Oysters

Oysters can generally be put into one of two groups: true oysters or pearl oysters. True oysters are the ones that people eat. Pearl oysters make balls that are commonly white, hard, and shiny, called pearls. People use pearls for jewelry.

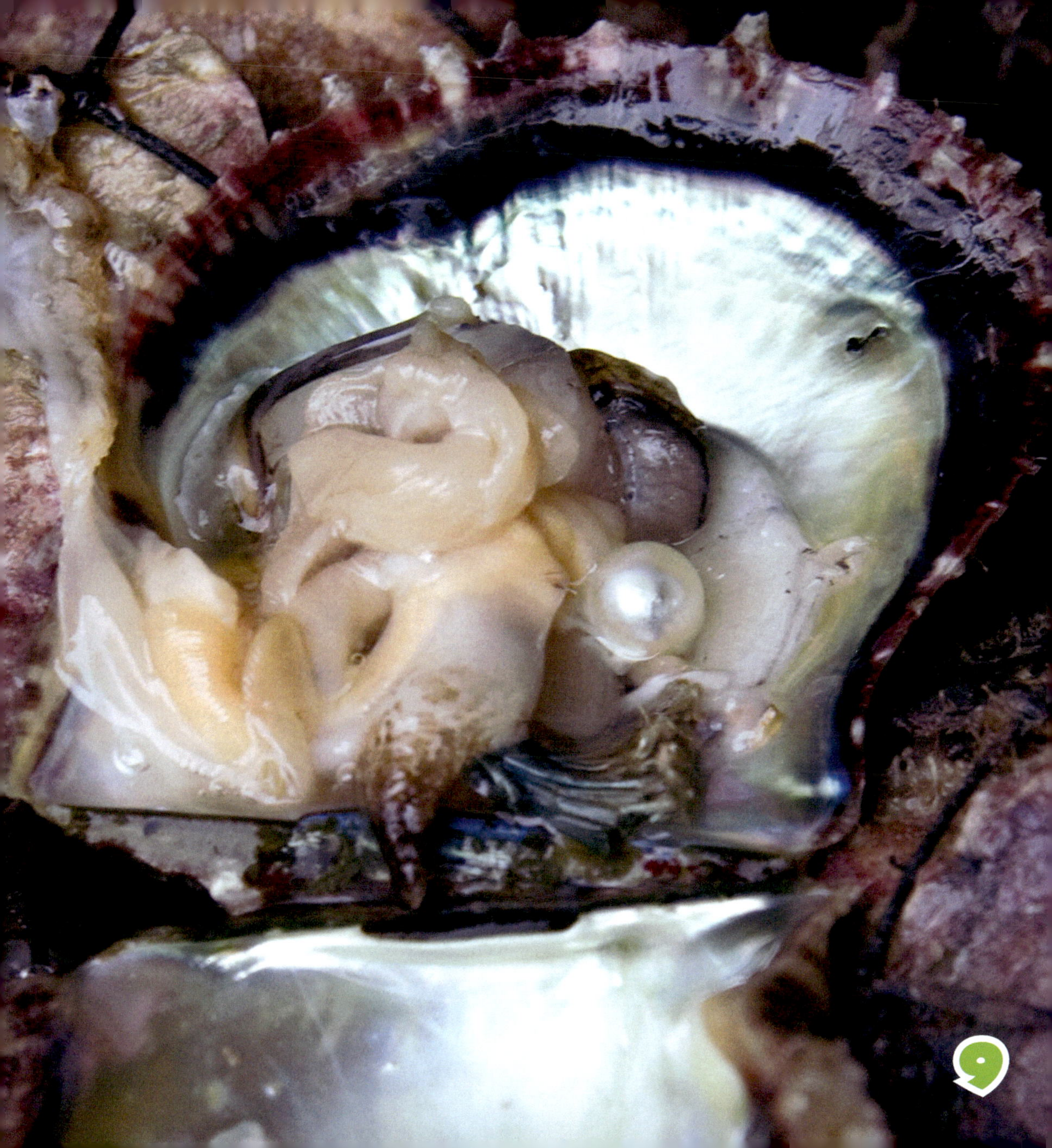

People and Oysters

In places along coasts, people commonly **harvest** true oysters. They're sold to stores and restaurants for people to eat. In addition to food, this provides jobs for many people. Oyster harvesting is good for the **economy** in these places.

Home for Many

People aren't the only ones who use oysters. Oyster reefs are key **habitats** for many ocean animals, such as **anemones**, crabs, and many types of fish. Animals may use oyster reefs to hide from predators or find food.

Keeping Things Clean

Oysters feed on **algae** and other bits that float in the water. As water passes through their **gills**, they trap anything they might want to eat. This is called filtering. It helps keep the water clean and makes for a healthier habitat.

Stopping the Storm

In places near coasts, storms and waves can cause serious harm to land. Oyster reefs help **protect** these areas by blocking waves and stopping some of the harm from storms. The reefs also protect underwater plants and animals.

We Need Oysters!

In some places, people have harvested too many oysters. This has caused the animals' **population** in these places to go down. In turn, the habitats the oysters created suffer. If oysters disappeared, many animals would be left without a safe or clean place to live.

Helping Oysters

Oysters are important parts of nature. When many are removed from water, more need to be added to keep the reefs and waters healthy. In some places, special programs, or plans, help control harvesting and keep oyster populations strong.

GLOSSARY

algae: plantlike living things that are mostly found in water

anemone: a small, brightly colored sea animal that sticks to surfaces and looks like a flower

attach: to fix to something

backbone: a row of connected bones that goes down the back of some animals

colony: a group of animals living and working together

economy: the money made in an area and how it is made

gill: the body part that ocean animals such as fish use to breathe in water

habitat: the natural place where an animal or plant lives

harvest: to bring in something, such as a crop, to be used

population: the number of animals of the same kind that live in a place

protect: to keep safe

FOR MORE INFORMATION

BOOKS

Kaufman, Alexander C. *Earth's Aquarium: Discover 15 Real-Life Water Worlds.* New York, NY: Magic Cat, 2021.

Morlock, Rachael. *Diving for Pearls.* New York, NY: PowerKids Press, 2018.

WEBSITES

Does Every Oyster Have a Pearl?
wonderopolis.org/wonder/does-every-oyster-have-a-pearl
Read more about oysters and the pearls they create.

Fun Facts About Oysters for Kids
sciencing.com/fun-oysters-kids-8371671.html
This website has more interesting facts about oysters.

Ocean Habitat
kids.nationalgeographic.com/nature/habitats/article/ocean
Learn more about ocean habitats and the plants and animals that call them home.

Publisher's note to educators and parents: Our editors have carefully reviewed these websites to ensure that they are suitable for students. Many websites change frequently, however, and we cannot guarantee that a site's future contents will continue to meet our high standards of quality and educational value. Be advised that students should be closely supervised whenever they access the internet.

INDEX

algae 14

anemones 12

crabs 12

filtering 14

fish 12

habitats 12, 14, 18

harvesting oysters 10, 18, 20

mollusks 4

oyster colonies 6

oyster reefs 6, 12, 16, 20

pearl oysters 8

pearls 8

plants 16

shells 4, 6

storms 16

true oysters 8, 10

waves 16